WHAT DOES AN
ANIMAL EAT?

WHAT DOES AN ANIMAL EAT?

By Lawrence F. Lowery

Illustrated by Bill Reusswig

NSTA Kids
National Science Teachers Association
Arlington, Virginia

National Science Teachers Association

Claire Reinburg, Director
Jennifer Horak, Managing Editor
Andrew Cooke, Senior Editor
Wendy Rubin, Associate Editor
Agnes Bannigan, Associate Editor
Amy America, Book Acquisitions Coordinator

ART AND DESIGN
Will Thomas Jr., Director
Joe Butera, Cover, Interior Design
Original illustrations by Bill Reusswig

PRINTING AND PRODUCTION
Catherine Lorrain, Director

NATIONAL SCIENCE TEACHERS ASSOCIATION
Gerald F. Wheeler, Executive Director
David Beacom, Publisher

1840 Wilson Blvd., Arlington, VA 22201
www.nsta.org/store
For customer service inquiries, please call 800-277-5300.

PERMISSIONS

Library of Congress Cataloging-in-Publication Data
Lowery, Lawrence F.
  What does an animal eat? / by Lawrence F. Lowery ; illustrated by Bill Reusswig.
    p. cm. -- (I wonder why)
  Originally published: New York : Holt, Rinehart and Winston, 1969.
  Audience: K to grade 6.
  ISBN 978-1-936959-46-4
  1. Animals--Food--Juvenile literature. I. Reusswig, William. II. Title.
  QL756.5.L68 2012
  664'.66--dc23
                        2012026553
eISBN 978-1-936959-62-4

Introduction

The *I Wonder Why* books are science books created specifically for young learners who are in their first years of school. The content for each book was chosen to be appropriate for youngsters who are beginning to construct knowledge of the world around them. These youngsters ask questions. They want to know about things. They are more curious than when they are a decade older. Research shows that science is these students' favorite subject when they enter school for the first time.

Science is both *what* we know and *how* we come to know it. What we know is the content knowledge that accumulates over time as scientists continue to explore the universe in which we live. How we come to know science is the set of thinking and reasoning processes humans use to get answers to the questions and inquiries in which we are engaged.

Scientists learn by observing, comparing, and organizing. So do children. These thinking processes are among several inquiry behaviors that enable us to find out about our world and how it works. Observing, comparing, and organizing are fundamental to the more advanced processes of relating, experimenting, and inferring.

The five books in this set of the *I Wonder Why* series focus on inquiry and various content topics: animal behavior, plant growth, physical characteristics of sound, animal adaptations, and mathematical measurement. Inquiry is a natural human attribute initiated by curiosity. When we don't know something about an area of our interest, we try to understand by asking questions and by doing. The five books are titled by questions children may ask: *How Does a Plant Grow? What Can an Animal Do? What Does an Animal Eat? What Makes Different Sounds?* and *How Tall Was Milton?* Children inquire about plants, animals, and other phenomena. Their curiosity leads them to ask about measurements, the growth of plants, the characteristics of sounds, what animals eat, and how animals behave. The inquiries lead the characters in the books and the reader to discover the need for standard measures, the characteristics of plant growth, sound, and animal adaptations.

Each book uses a different approach to take the reader through simple scientific information from a child's point of view: One book is a narrative, another is expository. One book uses poetry, another presents ideas through a fairy tale. In addition, the illustrations display different artistic styles to help convey information. Some art is fantasy, some realistic. Some art is bright and abstract, some pastel and whimsical. The combining of art, literary techniques, and scientific knowledge brings the content to the reader through several instructional avenues.

In addition, the content in these books correlates to criteria set forth by national standards. Often the content is woven into each book so that its presence is subtle but powerful. The science activities in the Parent/Teacher Handbook section within each book enable students to carry out their own investigations that relate to the content of the book. The materials needed for these activities are easily obtained, and the activities have been tested with youngsters to be sure they are age appropriate.

After students have completed a science activity, rereading or referring back to the book and talking about connections with the activity is a deepening experience that stabilizes the learning as a long-term memory.

All animals need food to live.

All animals will search to find food they can eat ...

wherever they can ...

whenever they are hungry.

Coyote

Roadrunner

Rattlesnake

Gopher

What kinds of food do you think these animals are looking for?

Some animals eat only plants.

Cows eat only plants.

Cow

Mountain Lion

Some animals eat only meat.

Mountain lions eat only meat.

Horse

Rabbit

Goat

Deer

Sheep

Squirrel

Here are some animals that eat only plants.

What are some other animals that eat only plants?

Plant eaters' front teeth are broad but sharp, like a spade. These front teeth are good for clipping off grass, leaves, and other plant parts.

The back teeth are large, square, and flat on top to provide a grinding surface. Plant eaters use these uniquely shaped teeth to crush and shred the food that was clipped from plants.

Coyote

Fox

Wild cat

Gray wolf

Here are some animals that eat only meat.

What are some other animals that eat only meat?

An animal that eats only meat has teeth that are a special shape.

The front teeth are sharp and pointed. Meat eaters use these teeth for biting, grasping, ripping, and tearing meat.

Meat eaters' back teeth are flat with jagged edges. Meat eaters use these specially shaped teeth to cut and shred the food they eat.

A cat has sharp, pointed teeth. So do lions, leopards, and tigers.

Tiger

A dog has sharp, pointed teeth. So do foxes, coyotes, and wolves.

A bear is an animal that can eat plants or meat.

Some of its teeth are sharp and jagged.

Some of its teeth are large and flat.

Grizzly brown bear

Raccoon

A bear also has long, sharp claws for capturing food.

A bear uses its claws when it goes fishing.

What shapes are your teeth?

Are you a plant eater or a meat eater, or can you be both?

Sparrow

Some animals have no teeth at all.
A bird is an animal that has no teeth.

Birds have tough, hard beaks. The shape
of a bird's beak can tell you about the
food it eats.

A sparrow has a short, thick beak. It is
good for crushing seeds. It is a seed eater.

Hawk

A hawk can rip and tear meat with its curved and pointed beak. It is a meat eater.

A pelican has a beak like a scoop.
It can scoop up and hold in its beak
the fish it eats.

White pelican

Crane

Some birds have long, pointed beaks.
These beaks are used to spear food.

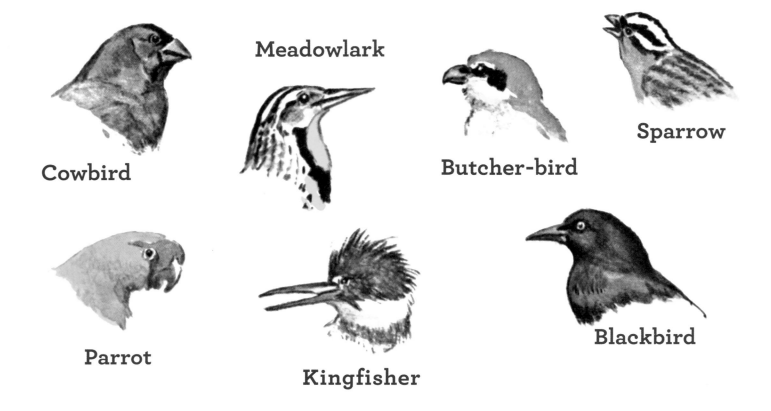

Cowbird

Meadowlark

Butcher-bird

Sparrow

Parrot

Kingfisher

Blackbird

Here are different birds with different beaks.
What kind of food do you think each bird eats?

Hummingbird

Duck

Woodpecker

Grouse

Jay

Egret

Robin

Toucan

Curlew

Rooster

Eagle

Owl

Hawk

An insect is another animal that has no teeth.

The grasshopper is an insect without teeth. Instead of teeth, it has many sharp moving parts in its mouth.

These parts move back and forth and from side to side to cut, chew, and grind the food that it eats.

Grasshopper

A grasshopper is a very small animal.
By itself, it does not eat many plants.

But now and then, grasshoppers come
in large swarms.

A swarm of insects can eat every plant on
a farm or in a field in a very short time.

Giraffe

If all the plants get eaten in an area, there will be no food left for the plant eaters.

The plant eaters will starve and die without food.

Lion

If most of the plant eaters die, there
will be no food for the meat eaters.

What will happen to the meat eaters?

When all the plants are gone, the plant-eating animals have nothing to eat. They starve and die.

When the plant-eating animals die, the meat-eating animals have nothing to eat. Then they, too, starve and die.

Zebra

Fortunately, birds eat many plant-eating insects. So it is rare that there are enough insects in any place at one time to eat all the plants.

This is one way that nature makes sure there is enough food for all animals.

Ivory-billed woodpecker

Robin

Sometimes people kill animals to protect food or other people. Sometimes people hunt animals. But people must be careful not to kill too many of any type of animal. What happens to one type of animal can affect the lives of other animals.

Some of the animals in this book are in danger of
being lost forever if people do not protect them:

the gray wolf   the cougar

the white pelican   the grizzly brown bear

the ivory-billed woodpecker

WHAT DOES AN ANIMAL EAT?

Parent/Teacher Handbook

Introduction

What Does An Animal Eat? is an expository text that presents how the structure of an animal's mouthparts provides clues to an animal's diet. The book also provides a simplified idea about food chains—that one animal can be the food source for another animal and, as such, is an integral part of the chain.

Inquiry Processes

This book engages the reader in making observations and comparisons of the mouthparts of animals (e.g., teeth, beaks, and mandibles) and the relationship of those structures to the types of food the animals eat. Questions are asked so that what is learned can lead to an understanding of the structure and function of human mouthparts and the role people can take to protect food chains.

Content

Knowledge of the structure and function of mouthparts sets a foundation for a deep understanding of (1) animal adaptations that contribute to their survival and (2) the far-reaching ecological relationships that bind together a seemingly incongruous assortment of plants and animals. One such relationship is called a food chain.

Food chains can range from a minimum of three levels to more than six levels. All the organisms in a food chain, except the first organism, are consumers. A simple food chain could start with three levels: grass, which is eaten by rabbits, which are eaten by foxes. A more complex food chain might begin with a butterfly that feeds on the nectar of flowers. The butterfly might be devoured by a dragonfly, the dragonfly by a bullfrog, the bullfrog by a water snake, and the water snake by a hawk. The food and energy are transferred from one organism to the next, linking them together and making them interdependent. If any set of organisms in the food sequence is depleted or exterminated, the rest of the chain will be affected.

Today, many people have little understanding of the part played by populations of animals and plants on the economy of nature. People may casually spray ponds with poisons to eradicate mosquitoes. The poisoned mosquitoes may be eaten by water organisms, which are eaten by fish, which are eaten by birds. The same people might then wonder why so many birds are found dead around the pond. The poisoned spray disrupted the food chain at one level and affected other levels.

Science Activities

1. Observe how some animals eat by placing them in a terrarium and providing them with appropriate food. For example, place a snail in a container covered with a piece of nylon stocking so the snail will not crawl out. Provide it with green leaves and observe the snail as it locates and eats the leaves. Pay special attention to the snail's mouthparts and how it uses them. Do this with other animals (e.g., grasshoppers, caterpillars, earthworms, hamsters, and so on). Feed them properly and observe what they eat, and when and how each feeds.

2. Take a trip to a zoo to observe different animals. Note the types of teeth or beaks of the animals. Try to determine what kinds of food the animals are fed at the zoo. Is the food appropriate for each animal based on its teeth or beak?

3. Collect pictures of animals from magazines. Paste the pictures into a chart to show the types of food each animal eats.

4. Use pictures from magazines to show a food chain.